# PRAISE FOR THE BRICK MATH
## *TEACHING MATH USING LEGO*

"I finally know what a fraction is. I can *see* it!"                    —Student

"Why doesn't everyone learn math this way?"                    —Student

"As an elementary teacher, exploring varying methods of learning is always necessary. From the very first activity in *Teaching Multiplication Using LEGO® Bricks*, it is clear that this book is extremely useful for any student learning (or struggling with) multiplication. For example, when learning/discussing fact families, I have witnessed many students blindly memorizing the facts without truly understanding *why* there is a relationship between the facts. By using different sizes of LEGO® bricks in one of the activities in this book, students are able to build and then observe a visual representation of the fact families. The students are able to see that one 1x6 brick contains the same number of studs as two 1x3 bricks.

In my experience as an educator, students tend to deeply grasp a concept whenever they are fully immersed in the learning process. The activities in this book require students to think critically about the process of multiplication that so often becomes robotic. *Teaching Multiplication Using LEGO® Bricks* covers multiplication processes such as: bundling, repeated addition, using place value, using array models, one-to-one correspondence, and more. Rather than blindly following a set of steps, students are able to build and think critically about what is happening as the problem evolves.

This book is a must-have for any educators exploring multiplication!"

—Elementary Teacher

"As an instructional coach at an elementary school, I have been searching for a teacher-friendly text that emphasizes the educational aspects of LEGO® bricks. *Teaching Multiplication Using LEGO® Bricks* helps breathe life back into mathematics, particularly multiplication instruction. The progression from basic multiplication principles to two- and three-digit multiplication problems is seamless. The students' understanding of these concepts is reinforced when using the LEGO® bricks, and the text encourages students to explain their findings. I recommend *Teaching Multiplication Using LEGO® Bricks* to everyone in education who wants to take the next step in hands-on learning."

— Kelli Coons, Instructional Coach

"*Teaching Fractions Using LEGO® Bricks* is a great resource for children to learn about fractions with conceptual understanding and modeling. It's hands-on, engaging, and overall an exciting way to learn about fractions. When you bring LEGO® bricks into the classroom the students automatically react with "oooh, cool!" and they are hooked on the activity. There is nothing better as a teacher than seeing your students enjoy learning, and using this resource, I see that. Another great feature about this resource is that it utilizes various learning modalities. Students learn physically by manipulating the LEGO® bricks, they draw the models for a visual reference, they write and describe concepts for a verbal understanding, and they are able to reason about the models and concepts to have a comprehensive understanding of fractions. Overall, this resource is phenomenal, and students are sure to be excited about math and fractions!"

—Tina Lupton, Teacher

"The visual models in *Teaching Fractions Using LEGO® Bricks* helped my students see and understand how equivalent fractions really work. The activities are super easy to follow and make learning operations with fractions fun for both the students and the teacher!"

— Jamie Piatt, Fifth Grade Teacher

Brick Math Series

# TEACHING DIVISION
## USING
## LEGO® BRICKS

Dr. Shirley Disseler

COMPASS

*Teaching Division Using LEGO® Bricks*

Brigantine Media/Compass Publishing
211 North Avenue
St. Johnsbury, Vermont 05819
Phone: 802-751-8802
Fax: 802-751-8804
E-mail: neil@brigantinemedia.com
Website: www.compasspublishing.org

ORDERING INFORMATION
**Quantity sales**
Special discounts for schools are available for quantity purchases of physical books and digital downloads. For information, contact Brigantine Media at the address shown above or visit www.compasspublishing.org.

**Individual sales**
Brigantine Media/Compass Publishing publications are available through most booksellers. They can also be ordered directly from the publisher.
Phone: 802-751-8802 | Fax: 802-751-8804
www.compasspublishing.org
ISBN 978-1-9384065-7-7

Printed in Canada

# CONTENTS

## DEDICATION

To my sons Steven and Ryan, whose love of LEGO® bricks as children inspired me to find ways to use the bricks in education to engage young minds in math!

And to the 2016 graduate students in STEM at High Point University for being the test audience for my activities!

## ACKNOWLEDGMENTS

Thanks to Neil Raphel and Janis Raye for their efforts in keeping me sane and moving forward throughout this project.

Much appreciation goes to Dr. Mariann Tillery, Dean of the School of Education at High Point University, for continuing support of all of my projects.

# INTRODUCTION

Division! Learning how to divide can be daunting for young students. Hours and hours of practice problems without a link to context is not the learning method that is supported in the math research. The best way to learn to divide is through modeling the relationship between the numbers.

The process of modeling the facts provides opportunity for the brain to utilize both creative processes and logical processes together. Research shows this to be the preferred learning format because it stimulates a need to know (Jensen 2005, Persaud 2013, Willis 2006).

The activities in this book provide practice with modeling the action of the two types of division (partitive and quotitive).

Jensen, Eric. 2005. *Teaching with the Brain in Mind, 2nd Edition*. Alexandria, VA: Association for Supervision and Curriculum Development.

Persaud, Ramona. 2013. "Education, the Brain, and the Common Core Standards." *Edutopia*. http://www.edutopia.org/blog/education-brain-common-core-ramona-persaud.

Willis, Judy. 2006. *Research-Based Strategies to Ignite Student Learning: Insights from a Neurologist and Classroom Teacher*. Alexandria, VA: Association for Supervision and Curriculum Development.

|  | **Partitive Division** | **Quotitive Division** |
|---|---|---|
| **Also called:** | • Equal sharing division | • Repeated subtraction division<br><br>• Measurement division |
| **What is known in the problem:** | • Total number of items being distributed<br><br>• Total number of groups getting the items | • Total number of items being distributed<br><br>• Total number of items in each group |
| **What is unknown in the problem:** | • The number of items per group | • The number of groups |

Starting with the basics of fact families, the activities in this book build in complexity. The final activities help students model division of two-digit and large numbers.

Typically, students learn these concepts in grades 3 through 6. Often multiplication and division are taught together to reinforce the relationship between them.

This book will help student master these topics, using a material found in almost every classroom and home— LEGO® bricks.

## Why use LEGO® bricks to learn about division?

LEGO® bricks help students learn mathematical concepts through modeling. If a student can model a math problem, and then be able to understand and explain the model, he or she will begin the computational process without struggling.

Modeling division with LEGO® bricks is an easy way for students to demonstrate understanding of the vocabulary and the concepts of whole numbers. When students model the action of division with LEGO® bricks, they have the opportunity to create multiple solutions for problems instead of looking for only one right answer. For division of whole numbers, the LEGO® brick also provides a way to practice and recall facts that is far superior to a rote memory technique.

LEGO® bricks are great tools for bringing many mathematical concepts to life: basic cardinality and counting, addition and subtraction, multiplication and division, fractions, data and measurement, and statistics and probability. Using LEGO® bricks fosters discussion, modeling, collaboration, and problem solving. These are the 21st century skills that will help students learn and be globally competitive.

The use of a common child's toy to do math provides a universal language for math. Children everywhere recognize this manipulative. It's fun to learn when you're using LEGO® bricks!

# USING A BRICK MATH JOURNAL

Journaling in math is an exciting way for students, teachers, and parents to review and share what is going on in the math classroom. A math journal is a resource that students can use for years to practice and review math concepts.

I recommend having your students start a Brick Math journal when you begin using LEGO® bricks to help teach math concepts. Here's how to use a Brick Math journal with the activities in this book: In each chapter, students begin in Part 1 (Show Them How) by building models that are teacher-directed. In Part 2 (Show What You Know), students build their own LEGO® brick models in response to specific prompts. Finally, students draw their LEGO® brick models on paper in their own Brick Math journals. The journal serves as a record of the physical models built that students can refer to over and over. The Brick Math journal can also serve as a form of assessment for teachers, a source for conferences, and as a way to identify if a student has any misconceptions in learning the topic.

Use these steps to create a Brick Math journal:

1. Use a composition book for each student. Set up pages one and two as the table of contents.

2. Have students number each page on the top outside corner.

3. Photocopy the base plate paper in the Appendix. Students will use the base plate paper to record their solutions, drawings, reflections, etc. Students will glue the base plate paper onto a journal page after they have drawn and colored their solutions.

4. When students glue in their drawings, they should label them with the title of the activity and then make the entry in the journal's table of contents.

.

# DIVISION FACTS

**Students will learn/discover:**
- The process of dividing with basic fact families
- What it means to divide parts of a whole

**Why is this important?**
Linking division to fact families is important when young learners begin to divide. The process helps them make sense of number relationships, opposite operations, and sets.

This activity uses a strategy called "stud covering" to show the quotient in a division model using bricks. Other activities in this book use different strategies. Students benefit by seeing a number of modeling strategies to find one that best suits their learning needs.

**Brick Math journal:**
After students build their models, have them draw the models on base plate paper and keep them in their Brick Math journals (see page 7 for more about the Brick Math journal). Recording the models on paper after building with the LEGO® bricks helps to reinforce the concepts and engages both the creative and logical thinking processes.

## SUGGESTED BRICKS

| Size | Number |
| --- | --- |
| 1x2 | 8 |
| 1x3 | 4 |
| 1x4 | 4 |
| 1x6 | 2 |
| 2x2 | 18 |
| 2x3 | 2 |
| 2x4 | 2 |
| 2x6 | 2 |
| 2x8 | 2 |

Note: Extra bricks are suggested for open-ended questions.

Note: Using a base plate will help keep the bricks in a uniform line. One base plate is suggested for these activities.

## Part 1: Show Them How

**1.** Place a 2x2 brick on a base plate. Count the studs (4).

Ask students to find two bricks that can be placed on top of the studs and cover them completely (two 1x2 bricks).

Explain that this model shows that 4 can be divided into 2 sets (the bricks), and each set has 2 studs.

Demonstrate how to write a division sentence for this model: $4 \div 2 = 2$.

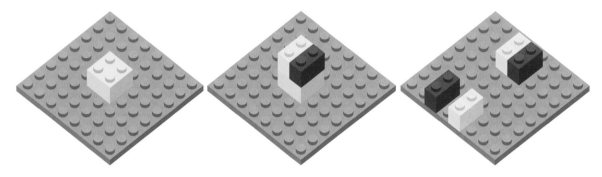

**2.** Place a 2x4 brick on a base plate. Count the studs (8).

Ask students to find four bricks that can placed on top of this brick with no studs uncovered (four 1x2 bricks).

Ask students what this model shows about these numbers.

(*Answer*: The model shows the number of sets is 4 and the number in each set is 2.)

Demonstrate how to write a division sentence for this model: $8 \div 4 = 2$.

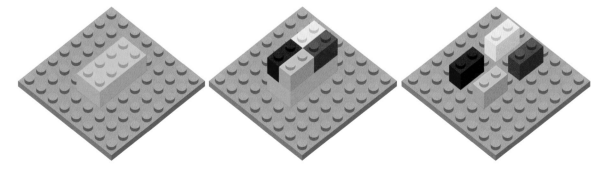

3. Place a 1x6 brick on a base plate. Count the studs (6).

Ask students to find three bricks that fit on top of this brick with no studs left over (three 1x2 bricks).

Ask students what this model shows. (Answer: 6 divided into 3 sets with 2 in each set.)

Demonstrate how to write a division sentence for this model: 6 ÷ 3 = 2.

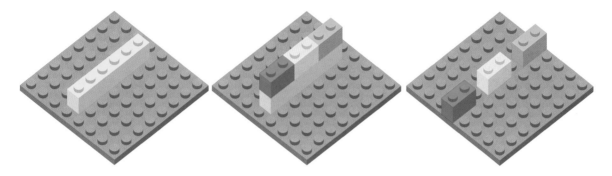

## Part 2: Show What You Know

**1.** Place a 2x4 brick on a base plate. What whole is being modeled? Find two bricks that fit on top without any left over studs. Decompose the whole into those sets. Draw your model. Write a sentence about your solution, using both words and a division sentence.

*Answer:* The whole is 8. The solution is 8 divided into 2 sets with 4 in each set; the division sentence is 8 ÷ 2 = 4.

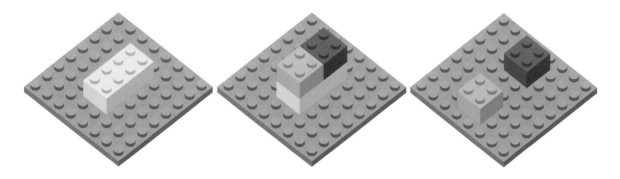

**2.** Place a 2x3 brick on a base plate. What whole is being modeled? Find three bricks that fit on top of the model without any studs left over. Draw and explain your model using both words and a division sentence.

*Answer:* The whole is 6. The solution is 6 is divided into 3 sets with 2 in each set; the division sentence is 6÷ 3 = 2.

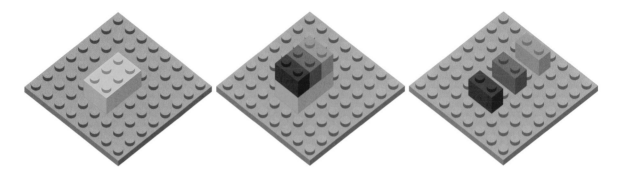

**3.** Model a whole with a brick of your choice. Decompose it into sets, showing two ways to divide it. For each way of dividing the whole, how many sets are there and how many are in each set? Draw and explain your solutions using both words and division sentences.

Solutions will vary.

# FINDING FACTORS

**Students will learn/discover:**
- What factors are
- How to find all the factors of numbers
- How to make models of factor families
- Vocabulary:
  - **Factors:** Factors are numbers you can multiply together to get another number. Example: 2 and 3 are factors of 6; 2 and 4 are factors of 8.

**Why is this important?**
Students need to be able to identify all the factors of numbers before they can work on equivalent fractions, simplifying fractions, and addition or subtraction of unlike denominators. For example: adding and subtracting fractions with unlike denominators requires a common denominator. Finding a common denominator requires knowing factors.

**Brick Math journal:**
After students build their models, have them draw the models on base plate paper and keep them in their Brick Math journals (see page 7 for more about the Brick Math journal). Recording the models on paper after building with the LEGO® bricks helps to reinforce the concepts.

## SUGGESTED BRICKS

| Size | Number |
|------|--------|
| 1x1 | 20 |
| 1x2 | 6-8 |
| 1x4 | 4-6 |
| 1x16 | 2 |
| 2x2 | 4-6 |
| 2x4 | 9-12 |
| 2x8 | 2 |

Note: Using a base plate will help keep the bricks in a uniform line. One base plate is suggested for these activities.

## Part 1: Show Them How
## Model how to find all the factors of 16

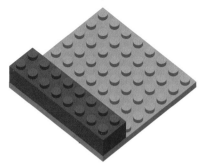

**1.** Place a 2x8 brick or a 1x16 brick on a base plate.

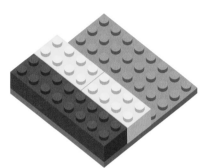

**2.** Place two bricks that are the same and, when placed next to the 16-stud brick, are equivalent in size and show two halves of the 16-stud brick. Use two 2x4 bricks or two 1x8 bricks.

**3.** Ask students: Can you find three bricks of equal size equivalent to the size of the 16-stud brick?

Let students look and think, and discover that the answer is no.

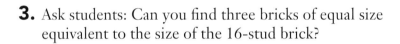

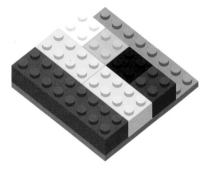

**4.** Ask students: Can you find four bricks of equal size equivalent to the size of the 16-stud brick?

Let students look and think, and discover that the answer is four 2x2 bricks or four 1x4 bricks.

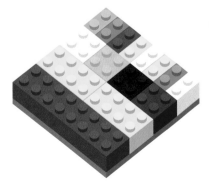

**5.** Ask students: Can you find the next number of equal-sized bricks that are equivalent to the size of the 16-stud brick?

Let students discover that five, six, and seven bricks don't work. Let them discover that the answer is eight 1x2 bricks.

**6.** Ask students: Can you find the next number of equal-sized bricks that are equivalent to the size of the 16-stud brick?

Let students discover that the answer is sixteen 1x1 bricks.

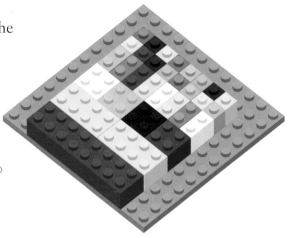

**7.** Name all the factors of 16 by looking at the LEGO® bricks on the base plate.

(*Answer:* 16, 8, 4, 2, and 1.)

## Part 2: Show What You Know

**1.** Can you build a model to show all the factors of 6?

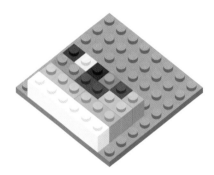

**Solution A:**
This model is a possible solution, showing factors 6, 3, 2, and 1.

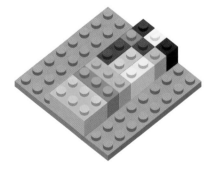

**Solution B:**
This model uses a different combination of bricks. Students who create this model could also explain that there are 2 sets of 3 in 6, and 3 sets of 2 in 6.

## Show What You Know

**2.** Can you build a model to show all the factors of 8?

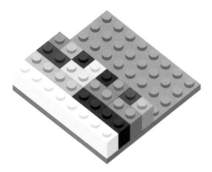

**Solution A**

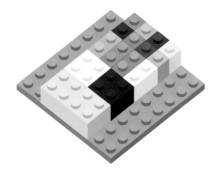

**Solution B**

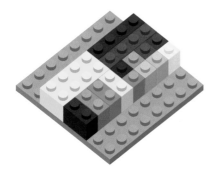

**Solution C**

Students who create models B or C could also explain that there are 2 sets of 4 in 8, and 4 sets of 2 in 8.

# EXPLORING DIVISION

**Students will learn/discover:**
- How division relates to multiplication
- Why division is actually repeated subtraction
- How to model division facts
- The vocabulary:
  - **Dividend** (the number being divided)
  - **Divisor** (the number of sets)
  - **Quotient** (the solution or answer)

**Why is this important?**

Students need to know what the term *divide* means. In order to have deeper understanding, students must recognize how division relates to both multiplication and repeated subtraction. This activity uses the strategy of set boxes.

Students will see that division is the opposite of multiplication by breaking down a whole into smaller sets when dividing. When multiplying, they put smaller sets together into a whole.

Students also will learn that, in the same way multiplication is repeated addition, one can think of division as repeated subtraction. When dividing, the same amount can be repeatedly subtracted from the whole until there are no more complete sets to subtract.

**Brick Math journal:**

After students build their models, have them draw the models in their Brick Math journals (see page 7 for more about the Brick Math journal). Recording the models in their journals after building with the LEGO® bricks helps to reinforce the concepts and engages both the creative and logical thinking processes.

## SUGGESTED BRICKS

| Size | Number |
|------|--------|
| 1x1 | 24 |
| 1x2 | 12 |
| 1x3 | 6 |
| 1x4 | 4 |
| 1x6 | 12 |
| 2x3 | 6 |
| 2x8 | 4 |

Note: A number of 1x10 or 1x12 bricks are also needed to serve as set separators.

Note: Using a base plate will help keep the bricks in a uniform line. One large base plate is suggested for these activities.

## Part 1: Show Them How

**1.** Build two set boxes on a large base plate using long bricks.

**2.** Ask students to think about the multiplication facts for 12.

**3.** Next to the set boxes, model the number 12 on a base plate using bricks of equal size (for example, six 1x2 bricks).

**4.** Ask students to count the studs. (*Answer:* 12)

Ask students how many set boxes are on the base plate. (*Answer:* 2)

**5.** Ask students to separate the 1x2 bricks into the set boxes evenly.

**6.** Ask students how many studs are in each block. (*Answer*: 6)

**7.** Discuss with students the vocabulary of the numbers.

Ask students: What do the 12 studs represent in the model? (*Answer*: the *dividend*, or the whole number being divided)

Ask students: What do the two set boxes represent in the model? (*Answer*: the *divisor*, or the number of sets the studs are being put into)

Ask students: What are the 6 studs called in each of the set boxes? (*Answer*: the *quotient*)

**8.** Discuss with students the terminology of a division problem. Explain that the *problem* is *12 divided by 2*, and the *solution* is *6 studs*.

**9.** Ask students: If we could write this problem as a multiplication problem, what would the problem look like and what would it mean? (*Answer*: 2 sets of 6 equal 12 or 2 x 6 = 12)

Ask students: Why is it not 6 x 2 = 12? (*Answer*: There are 2 sets with 6 studs in each set, not 6 sets with 2 studs in each set)

**10.** Show how division is like repeated subtraction.

Ask students: Is there another way to get the solution without grouping?

Show students the six 1x2 bricks. Allow students to speculate on another way to show the solution.

If needed, lead students by taking away one brick and then ask, "how many are left?" Make a table to help them understand the concept:

| Number of studs | Number removed (subtracted) | Number left |
|---|---|---|
| 12 | 2 studs removed (one 1x2 brick) | 10 |
| 10 | 2 studs removed (one 1x2 brick) | 8 |
| 8 | 2 removed (one 1x2 brick) | 6 |
| 6 | 2 removed (one 1x2 brick) | 4 |
| 4 | 2 removed (one 1x2 brick) | 2 |
| 2 | 2 removed (one 1x2 brick) | 0 |

Ask students, "How many times did you remove studs?" (*Answer*: 6)

Ask students, "How many studs were removed each time?" (*Answer*: 2)

Explain that this method is called *repeated subtraction*. It is another way to show that 12 divided by 2 equals 6.

**12.** Add a third set box to the model.

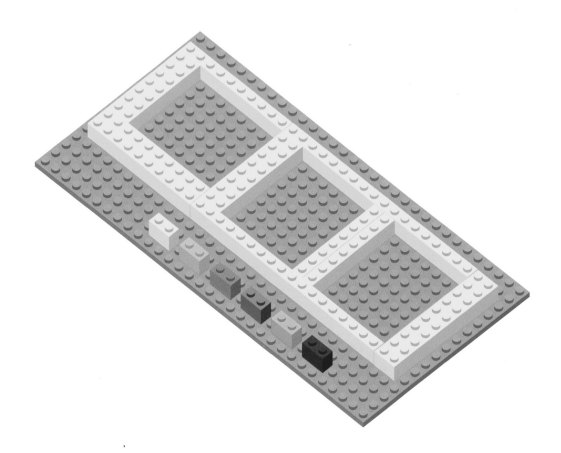

**13.** Use the same bricks (six 1x2 bricks) and separate them evenly into the three set boxes.

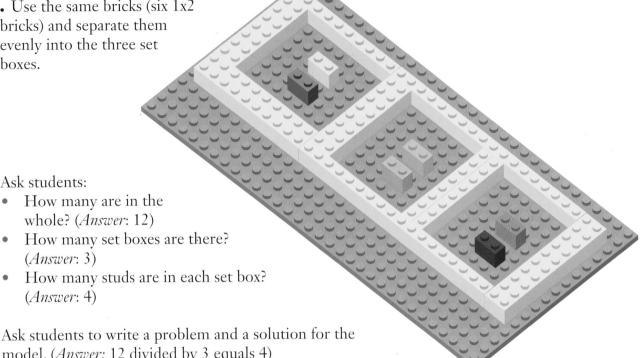

14. Ask students:
   - How many are in the whole? (*Answer*: 12)
   - How many set boxes are there? (*Answer*: 3)
   - How many studs are in each set box? (*Answer*: 4)

15. Ask students to write a problem and a solution for the model. (*Answer*: 12 divided by 3 equals 4)

16. Have students write a multiplication sentence for this problem to show the opposite operation. (*Answer*: 3 x 4 = 12)

17. Have students make a table to show the repeated subtraction that is taking place in the problem. *Answer*:

| Number of studs | Number removed (subtracted) | Number left |
|---|---|---|
| 12 | 4 studs removed (two 1x2 bricks) | 8 |
| 8 | 4 studs removed (two 1x2  bricks) | 4 |
| 4 | 4 studs removed (two 1x2 bricks) | 0 |

Have students explain how this chart shows the solution of 3 sets.

(*Answer*: Removing 4 studs represents one set of the model. Four studs are taken away three times to get to zero.)

*Challenge*: Make another model of division with twelve, using the same six 1x2 bricks. You can add or take away set boxes.

Possible solutions:

Students should discover that a four set box model does not work with the 1x2 bricks. This helps them understand the idea that decomposition is required to divide, which leads to the idea of remainders in later lessons.

Students might discover that since 12 is divisible by 4, they can decompose three of the 1x2 bricks to six 1x1 bricks, which will allow them to distribute the 12 studs equally by placing 3 into each of the 4 set boxes. The solution is 12 divided by 4 equals 3.

Another correct answer is to make a six set box model with one 1x2 brick in each block.

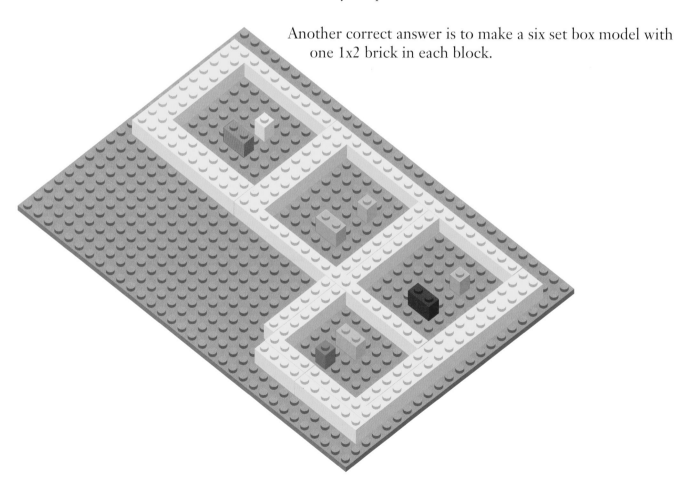

## Part 2: Show What You Know #1

**1.** Can you model and solve the division problem 24 ÷ 2 using the set box method?

Which number is the dividend and which number is the divisor?

*Answer*: The dividend is 24 and the divisor is 2.

Students should build a model with two set boxes. Students can choose to use several different bricks to represent 24: six 2x2 bricks, twelve 1x2 bricks, or eight 1x3 bricks.

Students should demonstrate the placement of these bricks in each of the set boxes.

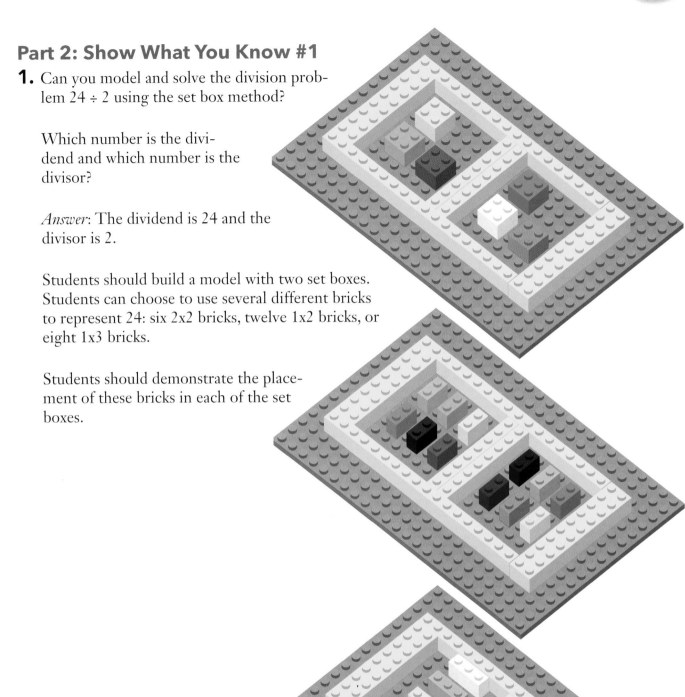

**2.** Can you write a multiplication sentence that shows the opposite of this division statement?

*Answer*: 2 x 12 = 24

**3.** What is the quotient, and how do you know?

*Answer*: The quotient is 12, because there are 12 studs showing in each of the two set boxes.

**4.** Can you draw a table to show the repeated subtraction that is taking place in the problem?

*Answer*:

| Number of studs | Number removed (subtracted) | Number left |
|---|---|---|
| 24 | 12 | 12 |
| 12 | 12 | 0 |

**5.** Record your model and the solutions to the problem in your Brick Math journal. Be sure to include your explanations.

## Show What You Know #2

**1.** Can you model and solve the division problem 15 ÷ 3 using the set box method?

Which number is the dividend and which number is the divisor?

*Answer*: The dividend is 15 and the divisor is 3.

Students should build a model with three set boxes. Students can choose to use two different bricks to represent 15: five 1x3 bricks or fifteen 1x1 bricks.

Students should demonstrate the placement of these bricks in each of the set boxes.

Students should place 5 studs in each set box. This will prove difficult if they choose to use 1x3 bricks. Students should realize they need to trade the 1x3 bricks for 1x1 bricks in order to evenly place them in the set boxes. This will show complete understanding and should be encouraged.

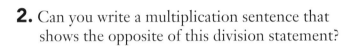

**2.** Can you write a multiplication sentence that shows the opposite of this division statement?

*Answer:* 3 x 5 = 15

**3.** What is the quotient and how do you know?

*Answer:* The quotient is 5. You know it because there are five studs in each box.

**4.** Can you draw a table to show the repeated subtraction that is taking place in the problem?

*Answer:*

| Number of studs | Number removed (subtracted) | Number left |
|---|---|---|
| 15 | 5 | 10 |
| 10 | 5 | 5 |
| 5 | 5 | 0 |

**5.** Record your model and the solutions to the problem in your Brick Math journal. Be sure to include your explanations.

## Show What You Know #3

**1.** Can you model and solve the division problem 18 ÷ 3 using the set box method?

Which number is the dividend and which number is the divisor?

*Answer*: The dividend is 18 and the divisor is 3.

Students should build a model with three set boxes to represent the divisor. Students can choose to use three different bricks to represent the dividend of 18: eighteen 1x1 bricks, six 1x3 bricks or three 2x3 bricks.

Students should demonstrate the placement of these bricks in each of the set boxes.

Students should place 6 studs in each set block: six 1x1 bricks in each set block, two 1x3 bricks in each set block, or one 2x3 brick in each set block.

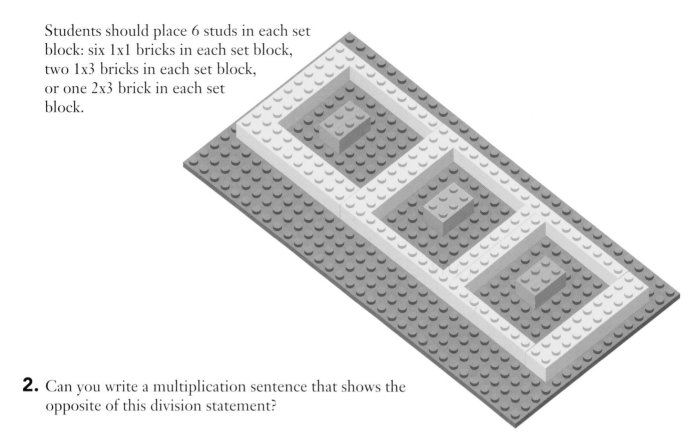

**2.** Can you write a multiplication sentence that shows the opposite of this division statement?

*Answer*: 3 x 6 = 18

**3.** What is the quotient and how do you know?

*Answer*: The quotient is 6. You know it because there are six studs in each box.

**4.** Can you draw a table to show the repeated subtraction that is taking place in the problem?

*Answer*:

| Number of studs | Number removed (subtracted) | Number left |
|---|---|---|
| 18 | 6 | 12 |
| 12 | 6 | 6 |
| 6 | 6 | 0 |

**5.** Record your model and the solutions to the problem in your Brick Math journal. Be sure to include your explanations.

## SUGGESTED BRICKS

| Size | Number |
|------|--------|
| 1x1 | 24 |
| 1x2 | 16 |
| 1x3 | 8 |
| 1x4 | 8 |
| 1x6 | 12 |
| 1x12 | 6 |
| 2x2 | 8 |
| 2x3 | 8 |
| 2x4 | 8 |
| 2x6 | 6 |

Note: Using a base plate will help keep the bricks in a uniform line. One large base plate is suggested for these activities.

# EQUAL SHARES OR PARTITIVE DIVISION

**Students will learn/discover:**
- The definition of partitive division
- What it means to have equal shares
- What it means to divide parts of a whole

## Why is this important?

Partitive division is a basic concept and involves taking the whole and dividing it into equal parts. Understanding what it means to divide into equal parts provides a basis for division as well as other math concepts such as fractions. Very early on, children begin to learn this idea when they are sharing.

This activity uses a different modeling strategy than has been used in earlier chapters. Students respond to different strategies depending on their individual learning styles, and it is helpful to expose them to a number of different strategies.

## Brick Math journal:

After students build their models, have them draw the models on base plate paper and keep them in their Brick Math journals (See page 7 for more about the Brick Math journal). Recording the models on paper after building with the LEGO® bricks helps to reinforce the concepts and engages both the creative and logical thinking processes.

## Part 1: Show Them How

**1.** Place a 2x6 brick on a base plate. Explain that this brick represents 12 pieces of candy—each stud is one piece. Tell students that will share this candy equally with a friend. Ask them to find two bricks that show how many pieces each person gets.

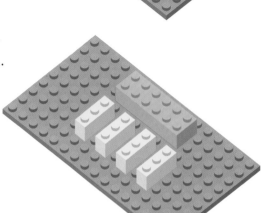

Students should choose two 2x3 bricks so that each friend gets 6 pieces of candy. Have students draw their solution model in their Brick Math journals and describe what the model shows.

**2.** Model how to share the same 12 pieces of candy among four friends. Ask students how that changes the solution.

Students should show four 1x3 bricks as the solution. Have them draw this model in their Brick Math journal and explain their solution.

**3.** Place a 2x3 brick on a base plate. Have students discuss some ways this amount could be equally shared among two friends, three friends, and six friends.

Introduce the term *partitive division*, explaining that it means dividing a whole into equal shares. Ask students to model the division, draw the model in their Brick Math journals, and write a word problem to show the equal sharing of the total amount.

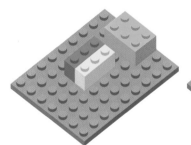
*Six candies shared equally between 2 friends.*

*Six candies shared equally among 3 friends.*

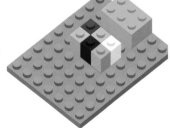
*Six candies shared equally among 6 friends.*

Note: Make sure students use the term "shared equally" in their word problems.

**4.** Place two 2x6 bricks on a base plate. Point out that this whole is equal to 24. Ask students to show as many ways as possible to equally share this amount, then draw their solutions in their Brick Math journals and write a word problem for each solution.

There are seven possible solutions. Four are illustrated:

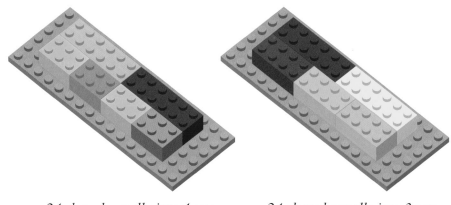

*24 shared equally into 4 sets of 6. Each one of the 4 friends gets 6 pieces.*

*24 shared equally into 3 sets of 8. Each one of the 3 friends gets 8 pieces.*

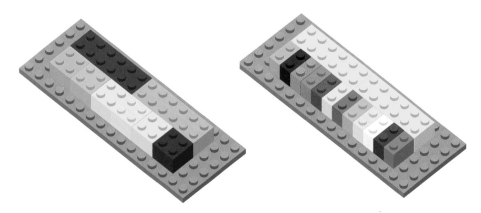

*24 shared equally among 6 friends. Each friend gets 4 pieces.*

*24 shared equally among 12 friends. Each friend gets 2 pieces.*

Correct models also include: eight 1x3 bricks, twenty-four 1x1 bricks, and two 2x6 bricks.

## Part 2: Show What You Know

**1.** Can you choose any brick that is at least 2x4 and show partitive division through equal sharing? Model your solution, then draw it and write a word problem for your model in your Brick Math journal.

*Answers will vary based on brick chosen.*

**2.** Place four 2x2 bricks together on a base plate. What is the divisor (the number being divided)? Can you show at least two ways to divide this number equally among groups? Model your solution, then draw it and write a word problem for your model in your Brick Math journal.

*Answers:* The divisor is 16. Students can use 2, 4, and 8 to divide the 16 studs equally. 1 and 16 also work, but usually there will not be enough 1x16 bricks available for students to model using this option.

**3.** Choose a number greater than 24. Model that number with bricks. Determine all the ways this amount can be divided equally. Draw all your models and write a word problem for each one in your Brick Math journal.

*Answers will vary.*

## 5

### SUGGESTED BRICKS

| Size | Number |
|------|--------|
| 1x1 | 24 |
| 1x2 | 16 |
| 1x3 | 8 |
| 1x4 | 8 |
| 1x6 | 12 |
| 1x12 | 6 |
| 2x2 | 16 |
| 2x3 | 10 |
| 2x4 | 8 |
| 2x6 | 3 |

Note: Using a base plate will help keep the bricks in a uniform line. One large base plate is suggested for these activities.

# REPEATED SUBTRACTION OR QUOTITIVE DIVISION

### Students will learn/discover:
- The definition of quotitive division
- What it means to use repeated subtraction

### Why is this important?
Using repeated subtraction to find out the number of groups in a whole helps with the foundations of algebraic thinking and the concept of missing variables.

### Brick Math journal:
After students build their models, have them draw the models on base plate paper and keep them in their Brick Math journals or workbooks (See page 7 for more about the Brick Math journal). Recording the models on paper after building with the LEGO® bricks helps to reinforce the concepts and engages both the creative and logical thinking processes.

## Part 1: Show Them How

**1.** Place a 2x10 brick on a base plate to represent 20 pieces of candy. This is the whole. If each person gets 4 pieces, how many people get candy?

**2.** Subtract 4 studs in succession to model the steps of this process. Introduce the term *quotitive division* and explain that it is repeated subtraction.

**Step 1:** a 2x10 brick as the whole of 20.

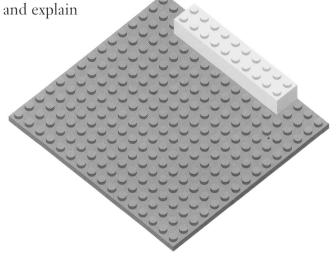

**Step 2:** a 2x8 brick that shows 4 less than the original whole of 20.

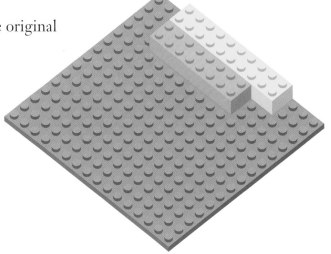

**Step 3:** a 2x6 brick that shows 4 less than the model of 16 in step 2.

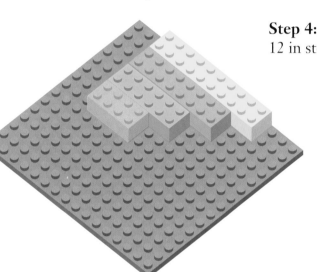

**Step 4:** a 2x4 brick that shows 4 less than the model of 12 in step 3.

**Step 5:** one 2x2 brick that shows 4 less than the model of 8 in step 4. If another set of four is removed, there would be zero studs remaining.

There are 5 steps in the repeated subtraction, so it shows that 20 divided by 4 pieces of candy means that 5 people can get equal amounts with none left over.

This model shows that $20 - 4 - 4 - 4 - 4 - 4 = 0$.

A division sentence for this problem is written as $20 \div 4 = 5$.

**2.** Place a 2x8 brick on a base plate. Ask students what whole is represented. (Answer: 16)

Ask students: if each person gets 2 equal shares, how many people can get an equal amount without any left over? Have students show the repeated subtraction model and count.

Each step shows the subtraction of 2 studs. Students should count the number of steps, including zero, for a total of 8.

Have students draw their model in their Brick Math journals, and explain using repeated subtraction vocabulary.

This model shows 16 – 2 – 2 – 2 – 2 – 2 – 2 – 2 – 2 = 0. A division sentence for this problem is written as 16 ÷ 2 = 8.

## Part 2: Show What You Know

**1.** Place a 2x4 brick at the top of the base plate. Count the number of studs in the whole. Can you determine how many equal groups of 2 there are in this model using repeated subtraction? Show the repeated subtraction problem and write the division sentence. Draw and explain your work in your Brick Math journal.

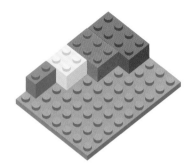

*Answer:*
This model shows 4 steps in the repeated subtraction. There are 4 groups of 2 that can be equally shared. The repeated subtraction problem is $8 - 2 - 2 - 2 - 2 = 0$.

**2.** Place two 2x6 bricks at the top of the base plate. What is the whole?

If there are 2 in each group, make a model to show the repeated subtraction. How many groups can have 4 parts of the model equally? Show the repeated subtraction problem and write the division sentence. Draw and explain your work in your Brick Math journal.

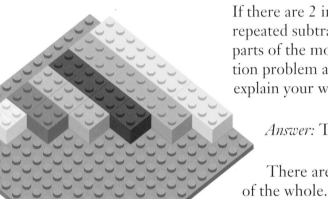

*Answer:* The whole is 24.

There are 6 groups that can share equally 4 parts of the whole.

The repeated subtraction problem is
$24 - 4 - 4 - 4 - 4 - 4 - 4 = 0$.

The division sentence is $24 \div 4 = 6$.

**3.** Choose a combination of bricks that shows a whole of 18. Can you build a model and explain two different repeated subtraction models for this whole? Can you write a repeated subtraction problem and a division sentence using this model? Draw and explain your work in your Brick Math journal.

*Answers will vary based on the repeated subtraction choice of each student.*

# TWO-DIGIT DIVISION

**Students will learn/discover:**
- The process of dividing a two-digit number by a one-digit number
- What it means to divide parts of a whole
- What it means to have a remainder

**Why is this important?**
Understanding the division of two-digit numbers provides a basis for mental math and estimation. This process is important for other concepts such as fractions and algebra.

This activity uses the modeling strategy of stud covering.

**Brick Math journal:**
After students build their models, have them draw the models on base plate paper and keep them in their Brick Math journals (see page 7 for more about the Brick Math journal). Recording the models on paper after building with the LEGO® bricks helps to reinforce the concepts and engages both the creative and logical thinking processes.

## Part 1: Show Them How

**1.** Model the number 24 with two 2x6 bricks of the same color. This is the dividend, or the whole that will be divided. Use 6 as the divisor (the number the whole is divided by). The division number sentence is 24 ÷ 6 = ☐

**2.** Ask students to find some bricks that have exactly 6 studs. Place the 6-stud bricks on top of the model of 24 until it is covered completely. Count the number of bricks covering the surface of the model of 24 to get the quotient (4). Have students draw the model and explain their thinking in their Brick Math journals.

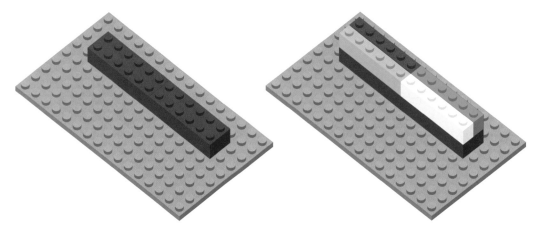

**3.** Build a model that shows 48 studs. This represents the dividend.

Cover the studs on this model using a divisor of 8. Have students write a division sentence for this problem. (Answer: 48 ÷ 8 = 6) Students could also view the model as 48 divided into 6 sets, with 8 in each set, or 48 ÷ 6 = 8.

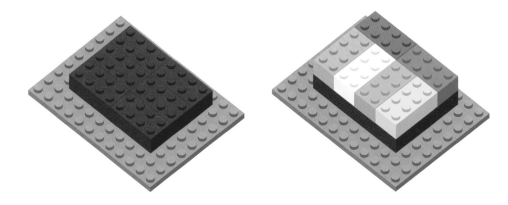

Have students draw their models and explain their thinking in their Brick Math journals.

**4.** Cover the same model with bricks to show 48 divided by 4. What is the quotient? (Answer: 12) What is the division sentence? (Answer: 48 ÷ 4 = 12) Students could also view the model as 48 divided into 12 sets, with 4 in each set, or 48 ÷ 12 = 4.

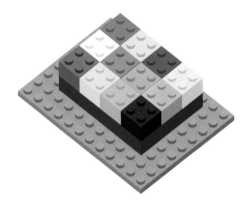

**5.** Model the number 18 as the dividend. Using 4 as the divisor, cover each set of 4 studs on the model.

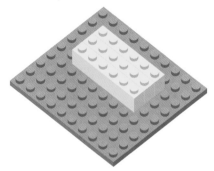
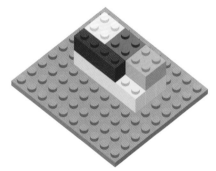

Ask students: what is different about this model? (Answer: There are 2 studs that cannot be covered by 4-stud bricks.)

Explain that the two studs left over represent the remainder.

The solution for this problem is written as: 3 bricks remainder 2 studs.

## Part 2: Show What You Know

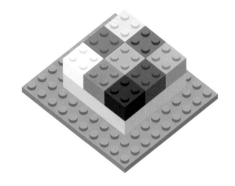

**1.** Can you model the number 36, then model this number divided by 4? Draw your model in your Brick Math journal and explain your quotient.

*Answer*: $36 \div 4 = 9$. Students could also view the model as 36 divided into 9 sets, with 4 in each set, or $36 \div 9 = 4$.

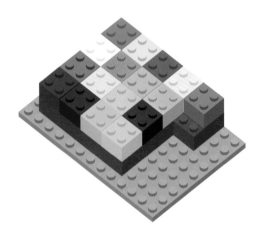

**2.** Can you model the number 70, then model this number divided by 4? What is your solution? Draw and explain your model in your Brick Math journal.

*Answer*: $70 \div 4 = 17$ bricks remainder 2 studs. Students could also view the model as 70 divided into 17 sets, with 4 in each set with a remainder of 2, or $70 \div 17 = 4$ bricks remainder 2 studs.

**3.** Choose any two-digit number and model that dividend. Find two numbers that divide this number equally without any remainders by placing bricks on top of the base model.

Show your solutions and write a division sentence for your models.

*Answers will vary.*

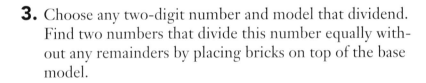

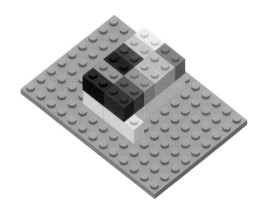

**4.** Model the number 25. Divide that number by 2. Draw your model and explain your solution.

*Answer*: Models should show twelve 1x2 bricks covering the bricks below them and 1 stud uncovered as the remainder.

## SUGGESTED BRICKS

| Size | Number |
|------|--------|
| 1x1 | 24 |
| 1x2 | 24 |
| 1x3 | 24 |
| 1x4 | 8 |

Note: Using a base plate will help keep the bricks in a uniform line. One large base plate is suggested for these activities.

# DIVIDING LARGER NUMBERS

## Students will learn/discover:
- How to use place value to divide with larger numbers
- The meaning of *remainder*

## Why is this important?
Students need to understand the concept of dividing larger numbers, not simply know how to perform the operations by rote. If they are able to relate larger numbers to place value, students improve their ability to do mental math and estimate. They learn to apply division to real-world uses for math, such as when purchasing items. For example, when purchasing a six-pack of soda for $3.50 and sharing the cost with 5 others, they will use division to find the cost per soda.

This activity uses a strategy of modeling with the bricks representing place value. This modeling technique is different from the techniques used earlier in this book for simpler division problems.

## Brick Math journal:
After students build their models, have them draw the models on base plate paper and keep them in their Brick Math journals (see page 7 for more about the Brick Math journal). Recording the models on paper after building with LEGO® bricks helps to reinforce the concepts and engages both the creative and logical thinking processes.

## Part 1: Show Them How #1

To model division of larger numbers, students must first relate the bricks to place value.

This modeling technique shows the ones place with a 1x1 brick, the tens place with a 1x2 brick, the hundreds place with a 1x3 brick, and the thousands place with a 1x4 brick.

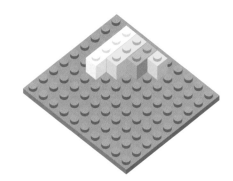

Have students build several numbers to practice using bricks to model place value. For example, model the number 2,345 with bricks representing place value.

When students understand how to model numbers using this place value method, use this technique to model division of larger numbers.

### Division of larger numbers (no remainder)

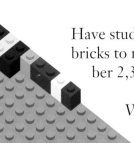

**1.** Build the number 222 using the place value method. Have students draw their models in their Brick Math journals.

**2.** Tell students that the divisor is 2 and that they need to find the quotient using the bricks to model the partitive division process of equal shares, with the number of groups unknown.

Ask students to write a division sentence for the problem. (*Answer:* The problem is 222 ÷ 2 = ☐.)

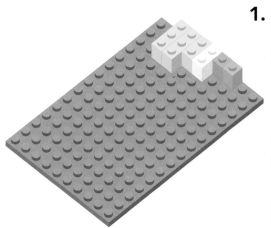

**3.** Place one 1x3 brick (representing 100) in each of two sets. Students can create set frames if they would like.

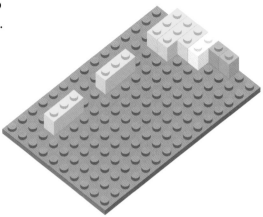

**4.** Divide the tens by placing one 1x2 brick (representing 10) in each set.

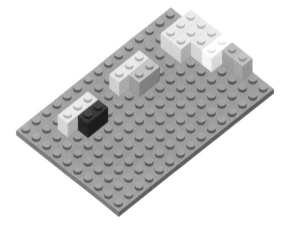

**5.** Distribute the 1x1 bricks (each representing 1) evenly into each set.

Each set contains 100 + 10 + 1 = 111, so the quotient is 111.

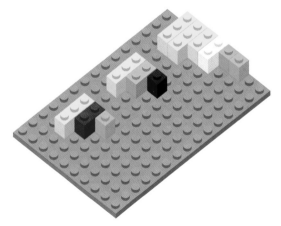

**6.** Have students draw their models and explain them in their Brick Math journals.

## Show Them How #2
## Division of larger numbers
## with remainder

$1{,}031 \div 2 = \square$

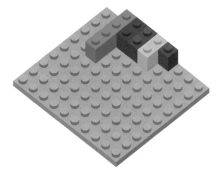

**1.** Model the number 1,031 as the dividend.

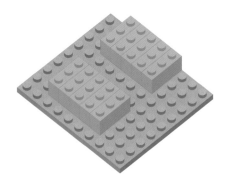

**2.** Use 2 as the divisor. Divide the 1000 evenly between 2 groups. The 1x4 brick must be decomposed into 10 1x3 bricks (representing 10 hundreds), and then 5 of each are distributed to each of the 2 sets.

(Note: If you do not have enough 1x3 bricks, students can use one 1x2 and one 1x1 to model one hundred.)

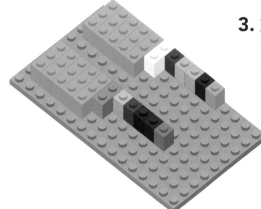

**3.** Divide the tens bricks into the two sets. Notice that there are 3 tens bricks and only 2 sets. Decompose one of the 1x2 bricks (tens) into 10 1x1 bricks (ones) and then divide them into the two sets with 5 in each set.

**4.** There is one 1x1 brick left over. This brick is the remainder and cannot be divided evenly without making it a decimal amount.

The quotient is: 515 remainder 1

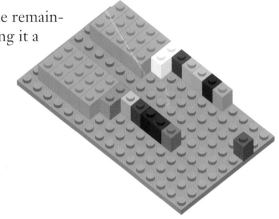

*More to learn:*

To show students how to find the decimal for the quotient, use a 1x2 brick as the whole, since the number of sets being used is 2. Place the one reminder brick on top of the 1x2 to show what decimal amount it represents. It shows that the 1x1 brick is ½ of the 1x2 brick. This means that the decimal in the quotient is ½. (Caution: This may cause confusion for students who are not able to transition from place value to decimal representations in the same model. This requires higher levels of understanding and ability to model.)

In decimal form, the quotient is 515.5

## Part 2: Show What You Know

**1.** Can you show a model for 444 ÷ 4 = ☐ ?
Draw and explain your model in your
Brick Math journal.

*Answer:* 111

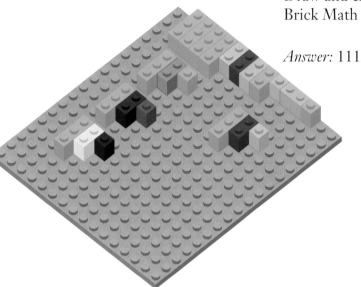

**2.** Can you show a model for 1,246 ÷ 2 = ☐ ?
Draw and explain your model in your
Brick Math journal.

*Answer:* 623

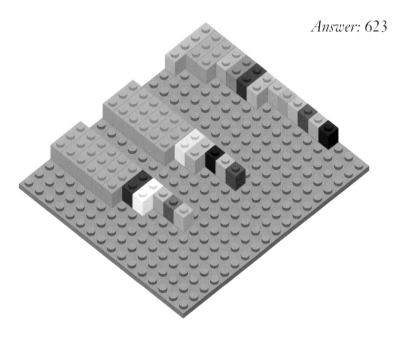

**3.** Can you show a model for 1,345 ÷ 2 = ☐ ?
Draw and explain your model in your brick journal.

*Answer:* 672.5 OR 672 remainder 1

*More to learn:*

Challenge: Show a model for 2,345 ÷ 3 = ☐

*Answer:* 781.67 OR 781 remainder 2

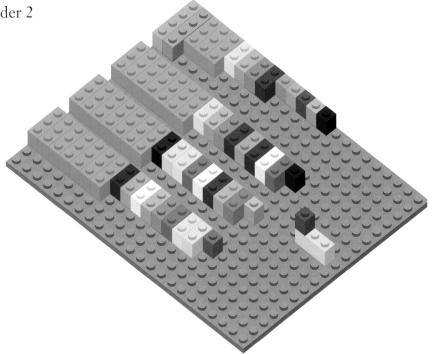

# APPENDIX

## Base Plate Paper

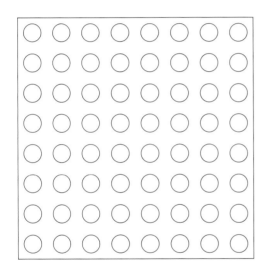

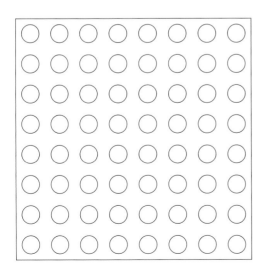

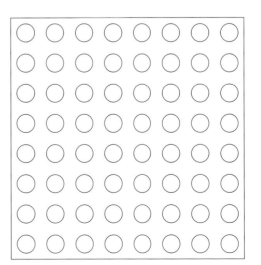

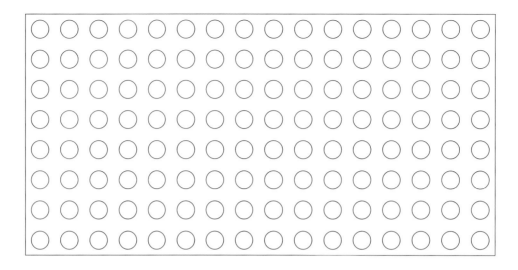

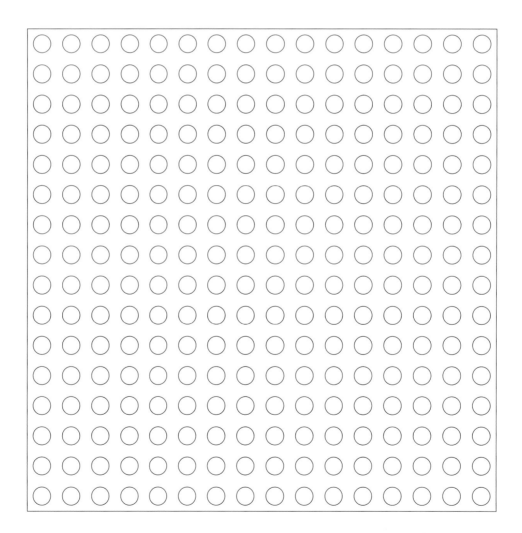

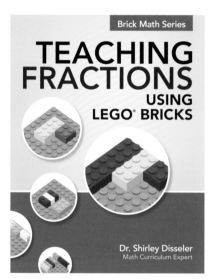

**Also in the Brick Math Series:**

# TEACHING MULTIPLICATION USING LEGO® BRICKS

Dr. Shirley Disseler

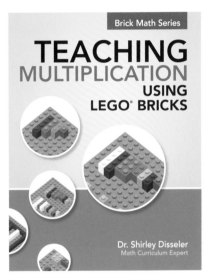

Teaching and learning multiplication is easy using LEGO® bricks!

Teachers as well as parents can follow the step-by-step instructions to guide students as they learn multiplication facts, one-digit multiplication, and two-digit and larger multiplication. Students model hands-on math problems with LEGO® bricks using a variety of techniques—sets, arrays, and place values—to develop true understanding of the concepts of multiplication.

Math is fun when you're using LEGO® bricks to learn!

Author Dr. Shirley Disseler is Associate Professor at High Point University and Chair of the Department of Elementary and Middle Grades Education. She serves on the LEGO® Education Ambassadors Panel.

**Companion student edition:**
## LEARNING MULTIPLICATION USING LEGO® BRICKS
Individual student book that follows the teacher's curriculum, complete with additional activities for practice and assessments.

---

Available on Amazon and at compasspublishing.org.

Quantity pricing and classroom packs available at 802-751-8802 or neil@compasspublishing.org.

COMPASS